FIRST EDITION.

Be

N

IMAGES OF THE WORLD

SEA

SAILORS

MARINE LIFE

BY Kaitlin Walkers

1663 Liberty Drive, Suite 200
Bloomington, Indiana 47403
(800) 839-8640
www.AuthorHouse.com

© 2005 Kaitlin Walkers. All Rights Reserved.

No part of this book may be reproduced, stored in a retrieval system, or transmitted by any means without the written permission of the author.

First published by AuthorHouse 09/29/05

ISBN: 1-4208-7234-6 (sc)

Printed in the United States of America
Bloomington, Indiana

This book is printed on acid-free paper.

Based on the works of
PAUL VALERY
Formerly printed by
FIRMIN-DIDOT PARIS
Numbered 1-500 Printed

English translation by Kate Walker
Photographic work by Colin Walker

PREFACE

The Sky and the Sea are the inseparable objects of the widest view; the most simple, the freest in appearance, the most changing in the entire extent of their immense unity; and nonetheless the most similar to themselves, the most visibly compelled to recover the same states of calm and torment, of trouble and clarity.

Idle, at the edge of the sea, if you try to decipher what is born in us in front of it; when, salt on your lips, and your ears gratified or offended by the murmuring or the roar of the water, you want to reply to this all-powerful presence, you have thoughts sketched out, scraps of poems, ghosts of action, hopes, menaces; a whole confusion of excited vague desires and images agitated by this grandeur which offers itself, which defends itself; which beckons venture by its surface, and frightens by its depths.

That is why there is nothing more insensitive, which has been more abundantly and more naturally 'personified' than the sea. We call it good, bad, treacherous, the fits, the drowsiness of a living being. It is almost impossible for the spirit not to naively love this great liquid body on which concurrent actions of the earth, the moon, the sun and the air makes up their effects. The idea of the capricious and violently voluntary character, which our forefathers attributed to their gods, and we ourselves sometimes attribute to women, quite imposes itself to that which borders the sea. A storm can suddenly erupt in two hours. A bank of fog thickens or clears as if by magic.

****

Two other ideas, too simple, and as if naked, are born again from the sea and from the spirit.

The one idea, 'fleeing for fleeing sake' which engenders a strange impetus of the horizon, a virtual momentum across the whole, a sort of passion or blind instinct from departure. The pungent smell of the sea, the salty wind that gives us the sensation of breathing from the area, the coloured and turbulent confusion of the harbours communicate a marvellous anxiety. The modern poets, from Keats to Mallarme, from Baudelaire to Rimbaud. Abound in impatient verses, which press the being and disturb it, like the fresh breeze through the rigging attracts the ships to anchorage.

The other idea is perhaps a deep cause of the first. You cannot wish to feel that which begins again. The infinite repetition, the rough and obstinate repetition, the monotonous crash and the identical resumption of the waves of the swell which sounds without respite against the limits of the sea, inspires the tired soul to foresee their invincible rhythm, the totally absurd idea of the "Eternal Return". But in the world of ideas, the absurdity does not hinder the power: the powerful and unbearable impression of an external restart changes into a raging desire to break the forever future cycle, irritates a thirst of 'strange froth', of virgin times and infinitely varied events......

****

As for myself, I sum up all this enchantment of the sea telling myself that it doesn't stop showing me the 'possible'. I spent hours watching it without seeing it, or observing it without a word inside. Sometimes I only get a universal image of it, each wave seems like a whole life to me. Sometimes now I only see what the eye naively experiences, and has no more. How do you detach yourself from such views? – Who can escape to the glamour of the living inertia of the mass of waters? It plays from transparency and reflections, of rest and movement, of peace and torment; disposes and develops in front of man, in fluid figures, law and fate, the disorder and the period; shows the route or paves the way.

****

A half-knowing, half-childish dream of the sea blurs, elucidates, combines a quality of memories or of spiritual wrecks of various orders and ages: childhood stories, memories of travel, navigational elements, fragments of exact knowledge...

Sometimes we know that this immense sea acts as a brake on the worlds, slowing the rotation. It is to the geologist the movement of a liquid rock, which holds atoms of all the bodies on the planets in suspension. Sometimes the spirit ventures into the depths. It feels the growing presence of it; it invents an increasingly gloomy dullness in it. Purer or warmer, or colder floods of water; internal rivers which circulate and close in on themselves in the mass; which divided themselves and join up again, lightly touching the continents, transports heat to the cold, brings cold back to the heat, creates ice hulls from the blocks, which break away from the polar ice fields – introducing a sort of exchange, similar to those of life, into the fullness and the continual substance of the inert water.

The very rapid vibrations, elsewhere, quite frequently agitate this great calm, quicker than sound, which cause underlying accidents, the abrupt distortion of the sea support. The muffled wave propelling itself from one extreme of the ocean to the other, suddenly collides with the monstrous platform of the risen-up lands, assails, crashes, devastates the populous platforms, and ruins cultures, homes and all life.

****

Where is the man who has not explored in spirit the abyssal nature? Just as there are famous sites that all travellers must have visited, there are fantastical places and imaginable states, which form in all minds, and artlessly, reply to the same irresistible curiosity.

We are all childlike poets when we dream of the bottom of the sea, and we lose ourselves there with delight. We create adventure and theatre in our minds by each imaginary step. Jules Verne is the Virgil who guides the young in this hell.

Slopes, plains, forests, volcanoes, desert pits, coral churches in half-living arms, luminous tribes, sprawling bushes, spiral creatures and scaled clouds, all this impenetrable and probable scenery is familiar to us. We circle in these coloured shadows, which change from liquid skies, where pass by moments, like bad angels of the sea, heavy and speedy forms of squalls on a cruise.

On the rock or in the silt, on a bed of shells or plants, sometimes comes gently, softly places itself, keels over, at the end of a slow descent, the enormous hull of a ship. There, under two thousand meters, the Titanic harbours a very complete collection of the material of our civilisation: the machines, the jewels, and the fashions of such a day.

****

But in the Oceans there are totally real and almost sensible marvels, about which the imagination is confused. I was speaking of under-water forests; that is to say a forest in a free state, without roots, denser, more entangled in itself, more teeming with life than the most virgin of terrestrial forests? Dream of this Atlantic region which encloses a loop of the Gulf Stream, and where the Sargasso sleeps, an immense mass of seaweed, a sort of obscure cellulose which only nourishes itself on the same water and enriches itself on all the bodies that this water holds dissolved. No ties fix on the depth of which the average depth is a league, this strange waterline, assembled over a space as vast as European Russia, and fabulously populated by all species of fish and crustaceans. Some authors evaluate the enormity of it, they say that it represents hundreds of millions of cubic kilometres of vegetable matter, in which are accumulated incalculable reserves of stalwart, potassium, chlorine, bromine, iodine, wrack.

This prodigious production of life, this mass of organic substance helps some spirits to understand the formation of oil deposits. The seaweed uncovered by the up thrust of the sea bed, little by little recovered and processed by rains, would decompose and reduce itself into hydrocarbons...

The sea is mysteriously linked to life, if, as so many people like to believe, life has a marine origin, one can note that in its first environment it shows itself infinitely more powerful, more abundant, more prolific than it is on land. Some areas of the sea, intermediate zones between the surface and the great depths, some variable routes through the shapeless water, are occupied or covered by unbelievable amounts of beings, sometimes the one pressed against the other more than one is in a crowd or on a roundabout in a capital city. Nothing makes you think about the true and naive nature of life moor than a shoal of fish. Perhaps, to express my sentiment, I should wish to write this word in singular - making "matter" out of these animals, composite matter, no doubt, of organised individual units; but of which the whole behaves like a substance submitted to very simple exterior conditions and laws.

I wonder of the price which we attach to existence, the value, the significance that we attribute to it, the metaphysical passion which we put into wishing that an individual is an isolable, incomparable event, produce once and for all. Is it not a sort of consequence of the rarity and mediocre richness of the mammals that we are?

We see in the sea that the extravagant multiplication of beasts which abound there is fortunately compensation by the destruction that they reek one against another. A hierarchy of eaters exists; and a statistical equilibrium ceaselessly re-establishes itself between 'eating species' and ' eaten species'.

So death seems to be an essential condition of life, and no longer an accident, which is a terrible marvel to us every time; it is for life, and no longer against it. 'To live' life must call to itself, 'breath in' so many beings each day, breathe out so many others; and fairly constant proportion must exist between these numbers. So life does not like survival.

Elsewhere, as to the degree of concentration of individuals which keeps a check on itself, in limited regions where life is most intense, it makes you dream of some property of the superficial liquid layer of the world, content of indistinct living beings in equilibrium with the state, the composition, the temperature, the movements of such a favourable area.

****

I believe the happiest beings of this world are found in a small group of porpoises, you see them from the top of a ship, and some believe you can see demi-gods in them.

Briefly mixed in foam, lightly touching the world of the air, playing with naked sun's fire; briefly even on the stern, battling with that which splits and divides the plenitude of the water, hunting down and covering the way just as dogs do in front of horses, they give the idea of fantasy in power. They are strong, alive, they have few fears; they move wonderfully next to the whole volume of their space, released from gravity, and free from all support: that is to say that they live in a state which we only dream about, and that we try to rejoin, awakened by the roundabout means of poisons or by the use of machines. The free mobility seems to man to be a supreme condition of "bliss", he pursues it endlessly, he simulates it by dance and music; he attributes it to the glorious bodies of the chosen. These leaping and diving porpoises offer him sight and inspire envy in him. That is why he also watches the ships, even the heaviest and ugliest, with all the interest in the means of movement in his heart.

****

It is not a delectable site - neither Alp nor Forest, nor monumental place, nor enchanted gardens -which- to my mind that which you can see from well-exposed terraces above the port. The eye knows inside out and at every hours of the day the sea, the town, their contrast, and all, which the ring broken by jetties and breakwaters encompasses, lets in and lets out. I breathe smoke, steam, perfume and breeze with delight. I love it all, even the straw and coal dust which rises from the quays; even the extraordinary smells of the docks and the warehouses where fruit, oil, livestock, green skins, fir tree planks, sulphur, coffees mix their factory values. I could spend days watching what Joseph Vernet called "the different jobs of a sea port" from the horizon up to distinct line of the constructed shore, and from the distant transparent coastal mountains to the ingenuous towers of semaphores and lighthouses, the eye embraces the human and inhuman are one at the same time. Isn't this the frontier where the eternally savage state, the brutally physical nature, the ever-primitive presence and the totally new reality meet, with the work of mans hands, the modified earth, imposed symmetries, arranged and created solids, displaced and frustrated energy, and all the apparatus of an effort, whose obvious law is purpose, thrift, suitability, expectation, hope?

Happy are the last ones resting on their elbows in the sun on the parapets of this so pure white stone of which the "Highways Department" build their sea walls and breakwaters! Others are stretched out on their stomachs on blocks, which are set forward that the incoming tide eats into cracks and disintegrates little by little.

Others fish; pricking their fingers under the water on the spines of the sea urchins, they attack the shells that are stuck to the rocks with knives. All around the ports there is a mob of such idle people, half-philosophers, half-molluscs? For a poet there are no companions who are more agreeable. They are the true amateurs of the marine theatre; none of the port life escapes them. For them, as for me, an entrance or an exit are always phenomena. You discuss the silhouettes spotted in the distance.

Some peculiarity in the forms or in the rigging generates theories. You judge the Captains character by the way in which the pilot who offers the service is welcomed.... But I'm no longer listening; what I see distances me from what they are saying. A large ship is approaching; a fisherman's vessel inflates the floats off towards the sea. The smoking enormity crosses the winged smallness in the channel, utters a strange bellowing, and drops anchor; the hawsehole suddenly vomiting its flow of links, with silvery noises, the thundering and high-pitched cries of a cast iron chain violently pulled from its shaft. Sometimes, as the rich brush against the poor in the street, the very pure, clean Yacht, all order and luxury, glides along the length of the vile coasters, small boats and ageless brigs full of bricks or barrels, loaded down with rusty things, of pumps which are in a sorry state, whose sails are in tatters; paintings, horrible accidents; passengers, chickens and a dog of uncertain breed. But it turns out that the ancient hull, which carries all these miserable creatures, is of a still beautiful line. Almost all the real beauties of a ship lie below the waterline; the rest is 'dead work'. Go into the holds or into the dry dock, consider the graces and forces of the hull, their volumes, the very delicate and minutely modulations of their forms, which must satisfy so many simultaneous conditions. Art intervenes here; there is no architecture more sensitive than that which finds a moving and driving ship at sea.

****

T*his presentation of old sea fairing photographs from a private collection of the author, portrays the ships, the seascapes, and the seamen, their dreams and their nightmares.*

SEA
SAILORS
MARINE LIFE

IMAGES OF THE WORLD

Sea Sailors Marine Life

Jibs, topsails and faces of prowess of half a century, it was again the dreams of the young officers of the Navy to command a Frigate. Here it is, moving vision of the ships of yesteryear. "The Melpomene" operating under her Latin veils. She was, until 1904 the school of military defence for the French Navy

The Elements of the sea, to the end of the journey, posses a particular style. The lifeless, melancholy anchor reflects in the calm waters of the harbour. Long chains dangle, and the sensual odour clears itself from the cockles and kelp.

To the shelter in the harbour, the boats of sin are united according to the sympathy of the skippers. The defects of the network, the ropes, the pulleys, and the sails detach themselves nervously, clean, the silhouette of a motor craft: The new race.

Kaitlin Walkers

Here are the riggings destined to stretch the sails that sustain the strength of a sailboat, these gracious attributes do not exist anymore.

The Three Masters spend eight months of the year at sea, and four months in harbour at St Malo, refitting and mending the sails, for preparation to return to sea.

Kaitlin Walkers

There are hundreds of rowboats similar to this one, and the sailors agree, that after a hard day, they would repair the nets burst by the porpoises and sharks, this they did in the harbour.

Here is one of the surface rocks that is marked on the charts by a small cross. The sailor will give the position of this rock to the navigator, and they will comfortably avoid it.

In the Breton Harbours the brown sails lowered is replaced by the unbroken nets that one has dried. The poets see in these immaterial things as a little dream that floats in the souls of the people of old Bretagne.

The outer harbour at Dunkirk

The flotilla of fishing boats arrange themselves in the outer harbour of Dunkirk. In the bottom of the picture the sailboats are uniformly linked together. Fast Manageable ships that one could overload with sails and whose manoeuvre was an art as much as a science.

Where the cross stands sentinel

    On the coast the surge never stops hitting. Here lie the tombs that don't contain coffins; some crosses are the symbols of all anonymous dramas.

Setting out for the banks of Flanders

The men leave for the fishing grounds of the banks of Flanders, the women look on from the end of the pier, they discuss the quality of the sailing, and discount what the catch might be, but hopeless departure is not without anguish

Work aboard Ship

Before entering into the regions of bad weather, the crew of the sailboats makes use of the time to strengthen the rigging. To rythmise the necessary efforts, the men sing the old sea songs.

In the grip of the storm

Crews are sent in the storm to tighten the foresail; they are ten meters above the bridge and are able to describe the sky, at every roll, of the impressive bows.

After the storm the ever-pounding sea has wrecked the sailboat, and the sea remains violent for several days. The men throw some of the cargo overboard to lighten the load, to avoid the lurking dangers.

The Storm bursts, and the sea attacks the decks and the riggings, the ropes snarl and the bare yards don't offer any hold to the wind. It can last for days or weeks.

Kaitlin Walkers

Undoing the work of the storm

Under the direction of the crews master the men in their shiny sowesters, execute the task, of untangling the ropes due to the very heavy winds.

The sailing ship in a tight place

The unfortunate sailboat is led to the trap. The tide, winds, will throw it against the rock, and the inevitable destruction will come true in a progressive way. The sails, the mast will fall, and soon the beautiful ship will be only wreckage along the coast.

Mid Atlantic Billows

The trails of froth on the sea distinctly indicate the sense of wind, we reduce speed, the spray shrouds the area like a cape, and the sea receives the ship like a blade cutting through, and through any shortcomings could make important damages.

Icebergs

In summer the fragments flow in enormous blocks that wonder until they reach moderate regions where they disappear, sculpted by their progressive melting, the icebergs take the most fanciful shapes, some would say like a marine monster.

Making ready to give up the ship

A reef? An iceberg? The wreck is imminent, and some crew endeavour to put the lifeboats to sea. The passengers help towards this task. They act quickly, but if the damage becomes more pronounced, the lifesaving will become impossible.

Sea Sailors Marine Life

Three-Master stranded on the rocks

Deceived by mist, the three-masted ship, expanded sails, came to break its stern against the rocks, the sea invaded the holds and the ship is now lost.

On the Newfoundland Banks

Not the least "Sheep" on the sea, very little wind, no wave, but a strong swell subsides with immense regularity creating a hollow often passing ten meters, where the rough trawlers nearly disappear completely.

Herring fishing

A good catch is discovered, the trawlers rush towards it, the nets are so full they overflow, on hoisting them; they loose part of the catch.

On the Lofoten islands one fishes the cod by nets or by the line. Here three men throw and bring in the lines without stopping while two others man the oars.

Sea Sailors Marine Life

Norwegian Fishing Boat
   The rough Norwegian sailors do not fear facing the storm on these half punt crafts, similar to the "Long ships" of their Viking forebears.

Kaitlin Walkers

Glaciers Seaward bound

Summer has come. The flow of ice parcelled itself out. The icebergs break and slip towards the sea. On this Nordic beach the seagulls assemble. A seal warms up in the sun, and one sees further up a lone penguin.

Sea Sailors Marine Life

Hauling the nets

One brings the nets on board and the traditional work commences. Fish accumulated on the deck from where they will be thrown into the hold.

Bringing in the Dragnets

    A wake of money. It is good for the" pocket" where after many hours the accumulation of fish is fruitful. Fishes deliver themselves from their ultimate fights.

The dogfish

Its not often that the luck of the fishing trip brings to the deck of the ship such hosts as this dog of the sea, of which the monumental proportions and skin are not unlike the stature and costume of the man that captured it.

Hoisting the loaded nets

The pocket of the dragnet is full and hoisted slowly aboard. When its contents have been poured onto the deck, fish will be sorted and quickly put into the hold.

Sea Sailors Marine Life

Cleaning the cod

Armed with a special knife, the fisherman disembowels the cod, after which the head is cut off and the central bone removed. Then he will be able to proceed with the salting.

Codfish heads

It is not so long ago that codfish heads were rejected and thrown back to sea. Today one preserves them and treats them industrially for the manufacture of manure.

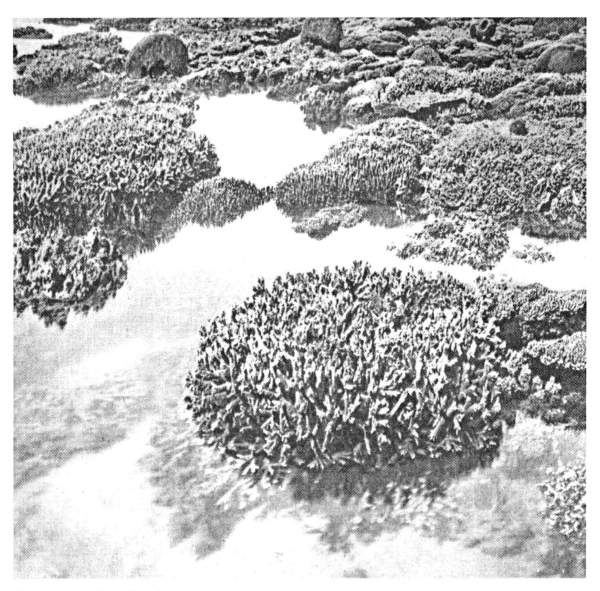

Great Australian Barrier

A barrier of coral sometimes surrounds the coasts of the Polynesian islands. Composed of a stony substance in the shape of plants, produced by millions of tiny animals.

This quiet water, this bay with the reflection of the palm trees in the water, it's the landscape, which marks the southern islands. Stevenson and Gauguins rocked their last dreams there, and it is here that is the pearl of the ocean: Tahiti.

Sea Sailors Marine Life

Penguins
The penguins live in the southern seas in groups whose organisation comes closer to the human societies; here they are going fishing.

Three-Master Close-Hauled

A rare meeting, of the beautiful "Square three-masted Ship" under all her sails, and the whaleboat passing very near, a pleasure for the passengers, and the sailboat benefiting from being at this precise point.

Sea Sailors Marine Life

The tails pierced by the harpoon, the Mammal is hoisted aboard the Whaler. Ships of 10 to 12,000 Tonnes that currently exist are real factories. The holds can extract all products from it, grease etc.

Kaitlin Walkers

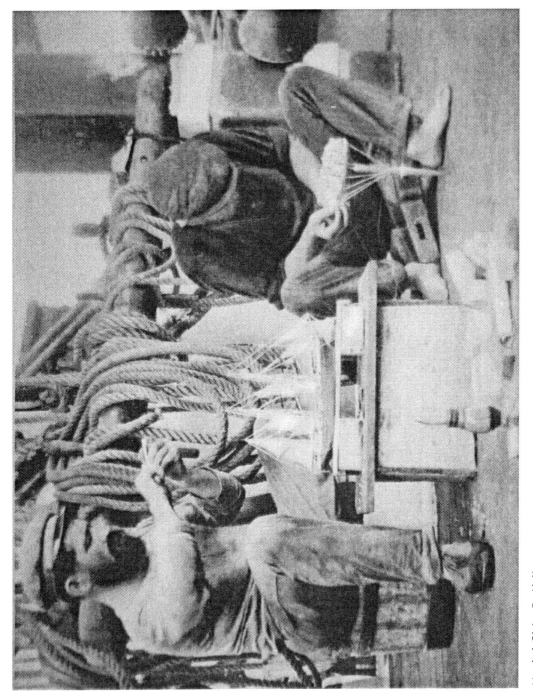

Model Ship Building
During the hours of rest, the sailors often construct with rudimentary means, models of the boats and ships in a bottle.

Sea Sailors Marine Life

Conger eel and Moray eel

 The Sailors diving the landscape see, battle ready, agitated Congers and Morays. The sailors pretend that the Moray attacked a man.

Barges of the Adriatic Sea

　　These Boats navigate the Adriatic, transporting heavy matters and fruits. Their rigs and sails are painted vivid colours as in the time of the Independence.

Sea Sailors Marine Life

Sea Flowers

   The gardens of the sea contain these flowers called zoophytes of which one exploited the shapes in decorative art.

Coral
  This old underwater landscape is made up of coral reefs.

On the coast of Belle island

The sea of Bretagne constantly agitates and develops the hardest granite. This full picture shows only needles, jagged rocks, and laces of stone.

Huge waves on the Jetty Head

When the large waves meet with the wind and spring tides, the seas fury reaches its severe sudden attack. Misfortune to the fishermen who didn't know it was time, and how to get back.

Vigil for the drowned sailors

In the Northern fishing harbours, it is custom devoutly preserved to make the vigil of the deaths for those that drowned at sea, and whose bodies are never recovered.

Old style Anchor

This anchor of centennial type, with its strong claws, clings on the seafloor allowing the ship to stop itself on the spot.

The Diver returning

The diver returns from the bottom, his diving suit is inflated with air under pressure. Before he unscrews his helmet he gives back his bag containing his tools to the aide onboard.

Mussel gathering

At the foot of the rocks, the low tide has shown mounds of Mussels. It is the work of the women to harvest these Mussels.

Sea Sailors Marine Life

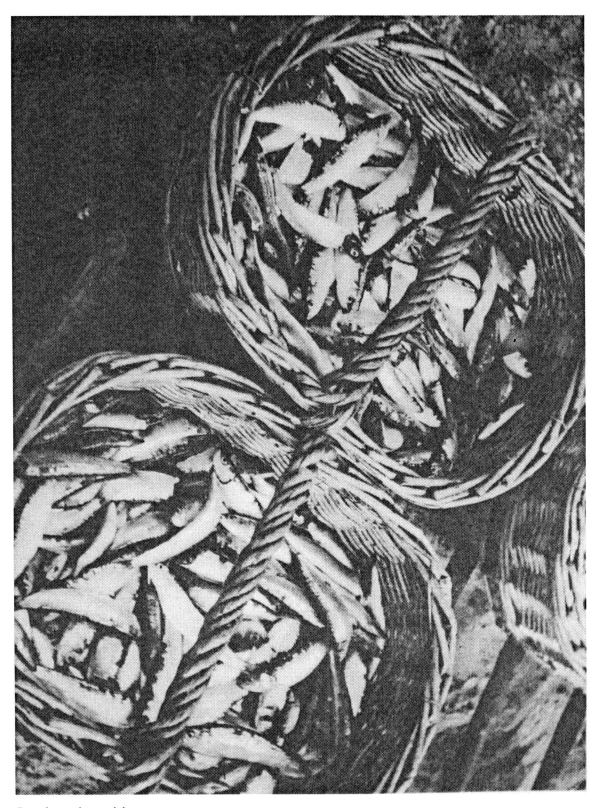

Ready to be sold
   The sardines and mackerel are sorted out by quantity and then shared into baskets, then carried to the fish stores or sold by auction.

Preparing the nets

When the mending is finished, the fishermen check very carefully the line of cork floats to insure that they will hold the nets vertically.

Dressing the Mackerel

Before putting the Mackerel into the baskets, the fishes are emptied washed and dried, plunged into boiling oil, and one will notice as luck will have it the fantastic shapes of the baskets.

Kaitlin Walkers

Fishermen cooking
   When the days fishing is complete, the nets are put out to dry, and the Bretons prepare their favourite dish made of fish and onions.

Stream between Ushant and the Continent

In the channel between Brest and Ushant, the currents are extraordinarily violent and the numerous dangers in spite of the perfection of the beaconing it is always necessary to fear the mist, calmness, or danger to shipping.

Kaitlin Walkers

Hoisting the spinnaker

  The double buoy, one hoists the "spinnaker" (on the left). A delicate manoeuvre but it increases the speed and she is well worth the short risk.

The Regatta

The disorder is only obvious, on the deck of the yacht that "let carry tall wind". Every man has his station and is ready to execute the manoeuvre ordered by the captain.

The behind wind

This beautiful pyramid of sails that transmits to the yacht, the effort of the wind is composed, while leaving the front of the jib, the big jib and the small jib, the mainsail looks more like an arrow.

Aloft to clear a rope

In a yacht race, the raising and lowering of sails, is done by ropes from the deck, but if a rope should tangle or break which could cost valuable time, the sailor must quickly climb to the top of the jib and clear the ropes, before loosing speed.

The end of the race is near
The competitors tighten up to reach the winning buoy, all get ready to transfer sails of which may depend on the result of the race.

Sea Sailors Marine Life

Cowes Regatta
The Cowes Regatta meets up with the huge ship "Rodney" the most powerful ship of war of all times.

French Cruiser "Tourville"

The "Tourville" that has for her tests has buckled the belt of the world, displaces 10,000 Tonnes. One endowed Cruiser with power of 120,000-horse power that allows her to cruise at 35 knots, she can sail close to the quarter of the terrestrial meridian without getting fresh supplies.

Submarines Requin and Espandon

The submarines Requin and Espandon have a displacement of 1.145 Tonnes on the surface and 1.440 submerged, their diesel engines give a speed of 16 knots on the surface, and 10 knots submerged, they can search 105 miles submerged at 5 knots and take thirty days cruising 9,000 miles at 9 knots.

French Airplane carrier cruiser
A plane is preparing to land on the deck of the carrier Le Bearn. One sees that the mast, bridge and funnel were constructed outside of the deck, to starboard, to clear the deck.

Sea Sailors Marine Life

Airplane Carrier cruiser
The navy's aircraft carriers constitute floating airfields, that follow the naval strengths and allow the take off and return of the hunting squadron, recognition and bombardment, cooperating with the land forces.

Airplane leaving the Aircraft carrier

Here one can see the deck of this aircraft carrier with groups of planes ready to take off, they group themselves in sections before taking the route to their objective.

Sea Sailors Marine Life

Attacking a shoal

The seagulls saw the shoal of fishes and came from afar, they hurry, and all you could hear were flapping of the wings, screaming, "lets dive", to take part in the feast.

Submarine coming to the mooring

Here is a Submarine called "Cotier" that is no heavier than 600 Tonnes on the surface (750 when submerged). Its speed and its staying power is lower than the bigger submarines, but it's arming of seven torpedo tubes makes it a dangerous contraption for fast operations.

Submarine running on the surface

The importance of the wake shows the escape of the smoke from the diesel engines behind and shows the submarine approaches at its maximum speed: 16 to 17 knots. The deck is nearly deserted; only 8 men remain in the conning tower.

Kaitlin Walkers

Under Brooklyn Bridge

The American Battleship Arkansas descends towards Liberty and passes under Brooklyn Bridge. One can distinguish the skyscrapers of Manhattan.

Battle practice

On the big ships of modern battle, methods of an unsuspected complication allow it to reach a simple solution. The "whole broadside" can be pointed and fired from a unique place in the topmast, here, in the distance a sailor has just fired. One can see the reflection of the flames on the water.

The front of a battleship turret

Above the powerful double turrets, the white bars are the telescopes; one distinguishes the spotlights on various decks. Finally the mast tripod carries on up to the lattice, which receives signals from the pavilion, and especially the eye of the ship, the station raised for the direction of firing.

An Electric crane

The modern ships of war are powerful fighting contraptions, but at the same time the ships endowed with lifting means, like this crane, they are made to lift several tons.

Kaitlin Walkers

Battleships heavy guns

These two tripled barrelled guns can be operated independently or all together from the central station, depending on the captains orders, the turrets are made of steel and are 457 millimetres thick. The structure of the guns is 356 mm. These guns are capable of firing 635kilogram shells 20,000 meters.

Sea Sailors Marine Life

Shrimp
Who thinks of the very small being when great weapons are sometimes used, this device of war is the one of the simple shrimp.

Launching of a torpedo

The torpedo merely "Fly's" from the aerial tube that one sees on the right, starts its motor and its mechanics of regulating organs (gyroscope immersion of regulating direction, etc). It spins towards its goal at 30 to 40 knots. Maximum range 10,000 meters. Loading of explosives 150 kilogram.

Depth charge exploding

The depth is the most powerful and efficient contraption in the struggle against the submarine. It is permitted to reach extreme depths, further than the submarine has dived. What damage it has caused, but what a magnificent flower of the water.

French Destroyer

The last construction of these, are of real small cruisers, 2,400 Tonnes, 5 canons of 138mm, 6 torpedo tubes, 8 officers and 300 men. Their speed is 36 knots (65km).

The crystalline coast
The crystalline rock that splits easily into layers on the coast of Arcachon protrudes into the sea like a prow of a ship, but the sea takes all: the rocks, the ships, the men.

Octopus
Thrashing right and left with its long tentacles, the octopus, accustomed to the seabed regains its den in the crevice of a rock.

Modern Diving Apparatus

Here is a perfect example of a diving suit established to reach depth of 100 meters, normally inaccessible until now. It was necessary to substitute the flexible costume of rubber with an extremely strong shell with slippery joints. The naked hands would not sustain the pressure of 10kg to cm2, so they replaced the naked hands with clamps and tools.

Destroyer raised at Scapa Flow
While one disputes more reduction of arming, a company undertook to recuperate the sunken armoury in Scapa Flow, nature took possession of this cannon, covered with algae and shellfish like a submarine rock.

Sea Sailors Marine Life

Air View of the Bremen

After a record crossing, the Bremen Ship arrives in New York. The tugs escort the ship to help her dock alongside the pier.

Unloaded ships

Contrast, barrels scattered and the beautiful order of ships, they are silent, disarmed, in for repair. A deaf rumour that emanates the distance, more contrast with embankments, chimneys and brilliant water.

Bologne Railroad and Navy station

The Navy and the trains desert the maritime station in the harbour of Bologne, by water or by rail, but along the embankment, the trawlers load their ships with ice and coal.

Coaling station
From the mines to the docks by barges and railroads, then from the docks to the lockers transported by skips, coal nourishes the majority of the ships again. In recent buildings the coal is replaced by fuel.

Air view of La Ciotat

The harbour of Ciotat specialises in the construction of transportation units for the lines in the Indies, Indo-China, Madagascar, and the oriental Mediterranean. One distinguishes a ship on the left in construction on hold. Another in the tip of the basin. In the harbour the other ships have either been constructed or waiting for repair.

Liner Il De France
The fastest and the most beautiful of the Liners. Here is the ultimate realization of the science of engineers. After a short voyage, the transatlantic Liner The IL De France moors itself at the embankment of the Havre.

Tugs' harbour in Hamburg
A big modern harbour, miles of embankments where the earth and the sea join, vast basins, and the auxiliary tugs are under pressure, as they are indispensable for the intense traffic.

Liners Engine room

The operations of the engine room could leave you dizzy, it's the clash of the beautiful, miles and miles of tubing, and this is where the mechanics work.

Sea Sailors Marine Life

D'IF CASTLE

Evoking the tragic adventures of Mira beau and the Count of Monte Cristo, the veiled fortress Chateau D'IF as seen from a plane, no doubt having been decorated again, and in the distance the mountains of the Estague reminds the painter Paul Cezanne of the familiar landscape.

Ship Building yard

The plates of sheet metal are nailed to the frame to form the shape of the ship. In a good few weeks one will be able to proceed with the launching. Thus under the immense crane of the rolling bridge, comes true the gestation of the gigantic ship.

Liners propellers

The giant like giant ship has been pushed onto a floating dock, the huge propellers looked like flowers on the end of "Trees". The propellers have four blades on each.

Crane

    To allow the ships the fast handling of the goods that they transport, the big harbours are provided with a big contraption for lifting, one sees here the Il De France being loaded in this way at the Havre.

Seaplane

The seaplane propelled by its motors, bounces in the slipstream before it takes off. Rather like a dragonfly scratching the agitated waves of a stream.

Liner coaling

It seems that one reaches here, the symbol of the maritime activity of our time. Produce of fire, coal, steel, unites to throw the navy towards the oceans adventure.

Sunset

    As the hour declines, whereas the seagull hovers for the last time in the overcast azure, the city dwellers on the ledge come to take the pure air.

Kaitlin Walkers

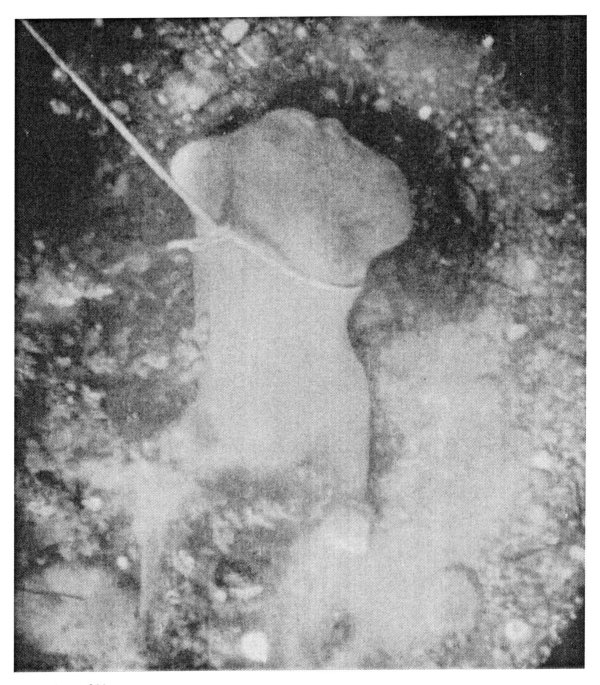

Emersion of Venus

To the large island of Rhodes, in the depths of the Aegean Sea a fisherman discovered laying on the seabed a big white body. Scientists came, embraced the torso of light. So the myth of yesteryear occurred again, the flower girl of the bitter wave.

Printed in the United Kingdom
by Lightning Source UK Ltd.
106855UKS00001B/9-18